Be a Water SCIENTIST

• Question • Experiment • Discover

By Ruth Owen

Ruby Tuesday Books

Published in 2024 by Ruby Tuesday Books Ltd.

Copyright © 2024 Ruby Tuesday Books Ltd.

All rights reserved. No part of this publication may be reproduced in whole or in part, stored in any retrieval system, or transmitted in any form or by any means, electronic, mechanical, photocopying, recording, or otherwise, without written permission from the publisher.

Editor: Mark J. Sachner
Design: Tammy West
Production: John Lingham

Photo credits:
Ruby Tuesday Books: 9, 14, 15T & C, 17, 18B, 19T, 20TC, 21, 23T, 25C; Shutterstock: Cover TL (Africa Studio), Cover TC (Kay Cee Lens and Footages/Yeti Studio), Cover TR (PeopleImages.com – Yuri A), Cover C (Khunkamo/Photodiem/GreSiStudio/Tomas Ragina), 1 (Illonajalll), 3 (Watson Images), 4TL (lisdiyanto suhardjo), 4TR (Theerasak Tammachuen), 4B (Scupix), 5T (fizkes), 5C (Krakenimages.com), 5B (EpicStockMedia), 6TL (PeopleImages.com – Yuri A), 6TR (VarnaK), 6B (Tatevosian Yana), 7, 8T (Illonajalll), 10T (Krisana Antharith), 10B (Anna Averianova), 11 (beebatch-photography/Paco Romero/Sinisa Botas/Mroald), 12T (Thong Xanh), 12BL (Labylullaby), 12BR (Krakenimages.com), 13, 14T (Yuliya Evstratenko), 14C (Lopolo), 15B (Mirvav), 16T (Dmitry Galaganov), 16C (Brian A Jackson), 18T (Uesiba), 18C (Denis Kuvaev), 19BL (Abdul Razak Latif), 19B (Retouch man), 20TL (Iurii Kachkovskyi), 20TR (bonchan), 20BL (rsooll), 20BR (Tomas Ragina), 22TL (New Africa), 22TC (PawelKacperek), 22TR (Billion Photos), 22B (Watson images), 23BR (Addictive Stock), 24T (Filip Fuxa), 24C (Ermak Oksana), 24B (xbrchx), 25TR (New Africa), 26T (Blue Cat Studio), 26B (KED 44), 27TL (PhotocechCZ), 27TR (Tatyana Vyc), 27CL (AaronChenPS2), 27B (Dmitry Rukhlenko), 30L (Littlekidmoment/SerPhoto), 30R (TriBayuMH/Vadim Verenitsyn/InFocus.ee), 31L (Maryna_L/Klymenok Olena/VH-studio), 31R (Byron Layton/HombreChicha/Samuel Borges Photography).

ISBN 978-1-78856-433-5

Printed in Poland by L&C Printing Group

www.rubytuesdaybooks.com

Contents

Our Watery World	4
Let's Get to Know Water	6
What Is Water Made Of?	8
Is Water Sticky?	10
Holding on Tight	12
Why Water Sticks	14
Where Did the Water Go?	16
From a Gas to a Liquid	18
Frozen!	20
Does It Dissolve?	22
Making Salty Seawater	24
Water in Action	26
Let's Talk Water	28
Glossary	30
Index	32

Our Watery World

Water is a big part of our world and daily lives.

Fresh water falls from the sky and flows out of taps.

Huge areas of our planet are covered with salty seawater.

Water inside you

More than half of your body weight is water! Where do you think all that water inside YOU is hiding?

(The answer is on page 28.)

You probably see, touch and taste water every day.

Now, it's time to investigate this wet, splashy stuff and. . .

. . .be a **Water Scientist**.

Let's Get to Know Water

Everything around us is made of **matter**. There are three states, or types, of matter.

Solids

A football and this book are both **solids**.

Liquids

Water is a **liquid**.

Gases

Take a big breath and blow. You can't see them, but you've just breathed in and blown out some gases.

Solids, liquids, and **gases** have **properties**.

A property is a quality that helps describe what an object or a substance is like.

What are the properties of water?

A good scientist asks lots of questions. They carefully observe things to see what they look like and to discover what they can do.

Use your science skills to observe water and discover its properties.

What words can you think of to describe water?

Write them in a notebook.

Take a jug of water and a glass. Pour the water back and forth between the two containers. What do you observe the water doing?

Does water have a shape? Look at these containers filled with coloured water. What do you observe about the water?

A liquid can be thick or thin. You can pour a thin liquid quickly from one container to another. A thick liquid is slow to pour. Do you think water is a thick or thin liquid?

Ketchup

Honey

Shampoo

Tea

Is water thicker or thinner than these liquids?

To find answers and more information, turn to page 28.

7

What Is Water Made Of?

To answer this question we need to learn some BIG science!

Everything in our world is made of tiny building blocks called atoms.

Atoms join together to make **molecules**. Water is made of molecules.

To make a water molecule, two hydrogen atoms and one oxygen atom join up.

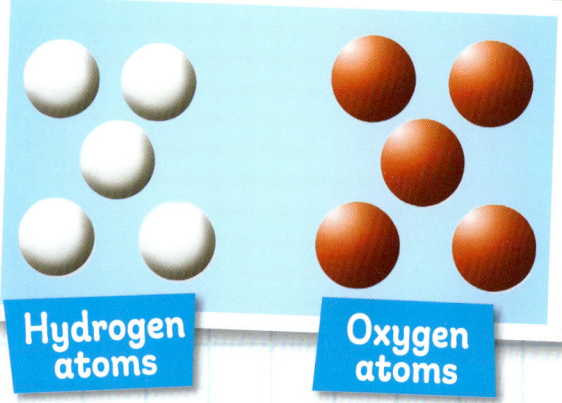

Hydrogen atoms

Oxygen atoms

Water molecule

Scientists draw a water molecule like this.

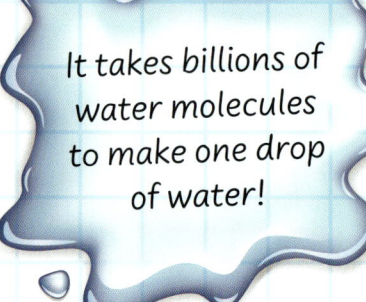

It takes billions of water molecules to make one drop of water!

Water drop

If we look at some water, we can't see the molecules. But sometimes we can see how they act.

You will need:
- 2 glasses
- Cold water
- An adult helper and some hot water
- 2 colours of food colouring

1) Fill one glass with cold water.

2) Ask your adult helper to fill the second glass with hot water.

3) Put a big drop of each food colour into the cold water.

4) Now carefully put a drop of each food colour into the hot water.

What do you observe the colours doing?

What do you notice about the way the colours are moving?

Remember! The water is made of molecules. What do you think the molecules are doing in the hot water?

To find answers and more information, turn to page 28.

Is Water Sticky?

The answer to that question is YES! However, water is not sticky like glue. It has its own special stickiness called **cohesion**.

If you spill some water, it forms a puddle.

That's because the water molecules are attracted to each other and stick together.

Puddle

Cohesion is one of the properties of water. It allows water molecules to form drops.

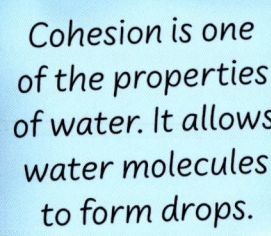

Water drops

How many water drops will stick to a coin?

You will need:
- A coin
- A cup of water
- An eyedropper
- A notebook and pencil

1) Place the coin on a flat surface.

2) Collect some water in the eyedropper. Carefully squeeze one drop onto the coin.

3) Now add a second drop next to the first.

What do the two drops do?

How many drops of water do you think you can fit on the coin?

4) Write your prediction in your notebook. Use tally marks to count the drops.

5) Keep adding drops to the coin until the water spills off the coin.

Try the activity again and see if you stick more water drops to the coin this time!

Holding on Tight

When you look at a water drop, it almost looks as if it has a skin — but it doesn't.

All the molecules in a water drop stick together because of cohesion.

But the molecules on the outside stick extra tight to the ones underneath them and on either side.

When water molecules act like this, it's called **surface tension**.

Surface tension

Have you ever over filled a glass of water? The water makes a scary bulge on top. But surface tension stops the water spilling over!

Let's investigate surface tension and how to break it!

You will need:
- A bowl of water
- A pepper grinder
- Some washing-up liquid

1) Fill the bowl with water.

2) Grind pepper into the bowl. Try to cover as much of the water's surface as possible.

What do you observe is happening with the pepper and water?

3) Now cover your fingertip with some washing-up liquid.

4) Gently touch the surface of the water in the centre of the bowl.

What do you observe the pepper doing now?

To find answers and more information, turn to page 28.

Why Water Sticks

Have you ever noticed how water drops seem to stick to other things?

Raindrops on a window

The molecules in the water drops are sticking together because of cohesion.

Then the molecules stick to other things because of **adhesion**.

When your skin is wet, it's because of adhesion. Water molecules are sticking, or holding tight, to your skin cells.

Hold on tight, everyone!
Water molecules
Skin cells

Can adhesion help water travel from one container to another? Let's investigate!

You will need:
- A ball of string
- Scissors
- A jug of water
- A small jug
- Duct tape (or other waterproof tape)
- A glass

1) Cut a length of string about as long as your arm. Soak the string in the jug of water for about one minute.

2) Tape one end of the wet string inside the small jug. Tape the other end inside the glass.

3) Fill the small jug with water.

4) Now gently lift and tip the small jug so the water slowly trickles out.

What do you observe happening?

How is it possible for the water to move in this way?

Adhesion is one of the properties of water. Drops of water are sticking to this spiderweb because of adhesion.

To find answers and more information, turn to page 29.

15

Where Did the Water Go?

When you climb from a swimming pool on a hot, sunny day, you might make a puddle or wet footprints.

However, the water soon dries up and disappears.

Where does the water go?

When we hang wet washing outdoors on a warm day, it gets dry.

The water doesn't disappear — it **evaporates**.

The Sun warms the water and makes it change from a liquid to a gas.

Let's test how water evaporates with this activity.

You will need:
- A jug of water
- Food colouring
- 2 jars that are the same size — one should have an air-tight lid
- A marker pen
- A notebook and pencil

1) Add food colouring to the water.

2) Fill each jar three-quarters full with water. The water level in each jar should be exactly the same.

3) Mark the level of the water in each jar with the marker pen.

4) Tightly screw a lid on one of the jars. Stand both jars on a hot, sunny, indoor windowsill.

5) Every few days, check your jars. If the water level changes in a jar, mark the level on the glass.

What do you predict will happen to the water? Write your prediction in your notebook.

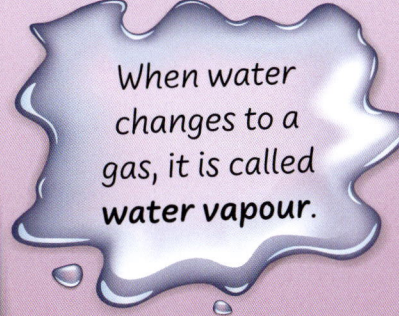

When water changes to a gas, it is called **water vapour**.

What do you observe happening?

In your notebook, write a sentence that explains your results.

From a Gas to a Liquid

When we use a hairdryer, the liquid water that's sticking to our hair evaporates.

The water becomes water vapour, and our hair feels dry.

Water vapour is invisible. We can't see it floating in the air, but it's there.

Can water vapour become liquid water again? Yes!

When water vapour cools down, it **condenses** and changes into liquid water.

Water vapour → Cold air → Liquid water drops

Is there water vapour in the air around you right now? Let's investigate.

1) Stand two glasses on a table or counter.

You will need:
- 2 glasses
- 5 ice cubes
- A phone or watch for timing the experiment

2) Put five ice cubes in one of the glasses.

3) Wait for 15 minutes.

What do you notice about the glass that contains the ice cubes? How is it different from the other glass?

What do you think has happened?

Take a can of drink from a fridge. Water drops will soon appear on the can. Why?

To find answers and more information, turn to page 29.

Frozen!

Remember how we learned there are three states of matter — liquid, gas and solid?

Water is amazing stuff because it can be all three!

Liquid water

Water vapour gas

Solid ice

When the air temperature goes down to 0° Celsius, liquid water freezes.

Frozen lake water

The water changes into solid ice.

We measure temperature with a thermometer. This thermometer shows freezing point.

When ice gets warm, it melts and changes back to liquid water.

Can we melt ice without heat? Let's investigate!

You will need:
- 4 small bowls
- 12 ice cubes
- 2 tablespoons of salt
- 2 tablespoons of sand
- 2 tablespoons of sugar
- A notebook and pencil
- A phone or watch for timing

1) Put three ice cubes in each bowl.

2) Quickly sprinkle salt over one bowl of ice cubes. Sprinkle sand over a second bowl and sugar over a third.

Salt **Sand** **Sugar**

3) The fourth bowl will just contain ice cubes.

4) Place all four bowls in a fridge.

Which substance do you think will melt ice the fastest?

5) Predict the order in which the bowls of ice will melt. Write your predictions in your notebook.

6) Check the bowls of ice cubes every 20 minutes for one hour.

Liquid water doesn't have a shape. But ice is a solid, so it forms shapes.

To find answers and more information, turn to page 29.

21

Does It Dissolve?

When water is in its liquid state, it's possible to mix it with other things.

Let's take a glass of water and stir in a teaspoon of sugar.

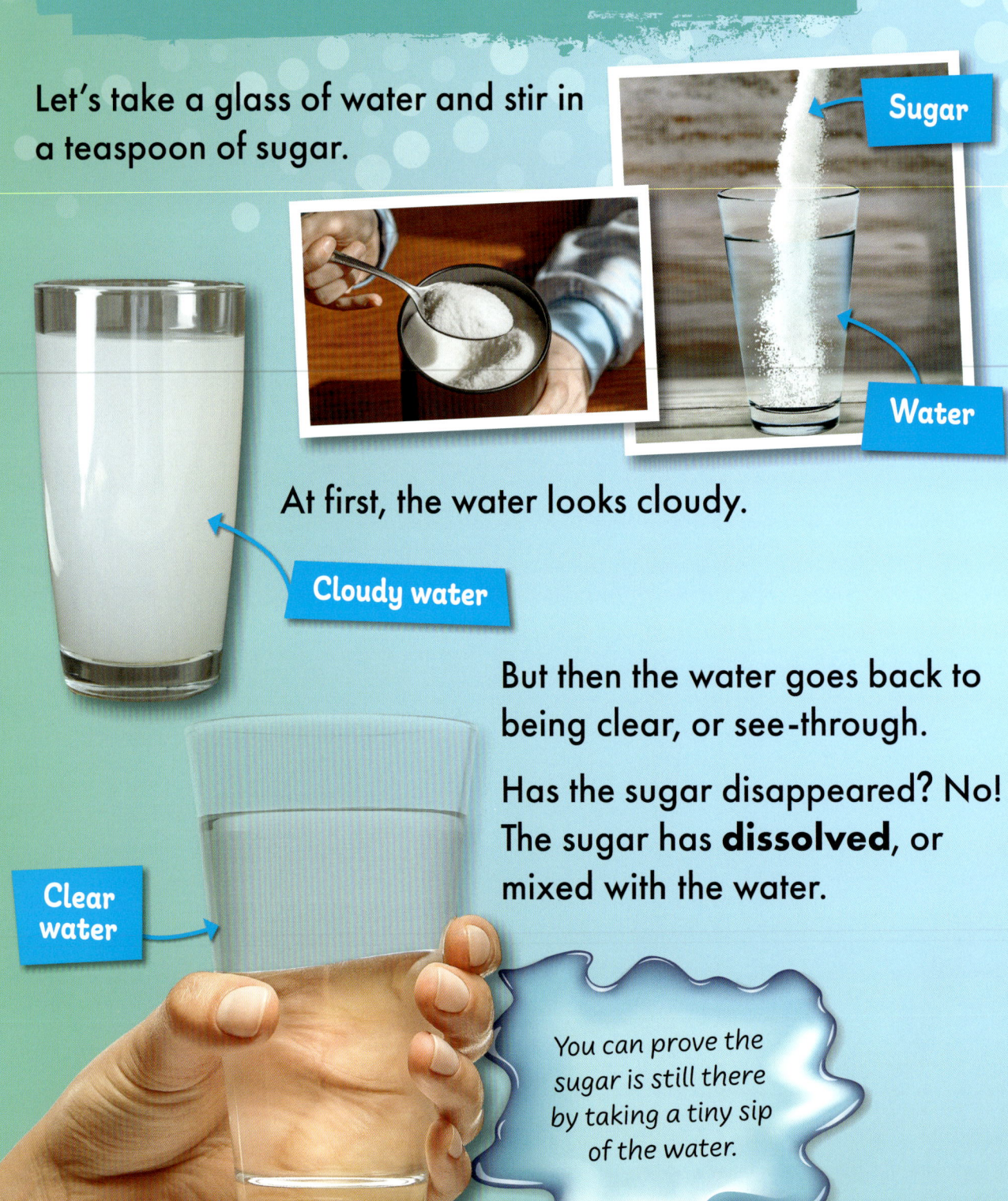

Sugar

Water

At first, the water looks cloudy.

Cloudy water

But then the water goes back to being clear, or see-through.

Has the sugar disappeared? No! The sugar has **dissolved**, or mixed with the water.

Clear water

You can prove the sugar is still there by taking a tiny sip of the water.

What other substances will dissolve in water? Let's test some everyday things!

You will need:
- Substances for testing, such as the ideas here or your own choices
- Small glasses, jars or recycled, clear plastic containers
- A jug of cold water
- A teaspoon
- Kitchen towels for wiping the spoon
- A notebook and pencil

1) Examine each substance by touching it and carefully looking at it.

Ground pepper

Sand

Chalk

Salt

Instant coffee

Flour

Tea
(from inside a teabag)

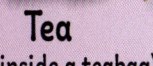

Dried herbs

Turmeric

Predict what will happen when you mix each substance with water.
- It will dissolve.
- It will float and won't dissolve.
- It will sink and won't dissolve.

Write your predictions in your notebook.

2) Fill a container with water. Stir in a teaspoon of one of the substances. Observe what happens.

3) Record your results in your notebook.

4) Repeat steps 2 and 3 with the other substances.

Did your results match your predictions?

Can you tell if a substance will dissolve in water by feeling it?

Making Salty Seawater

If you've accidentally swallowed some ocean water, you will know it tastes salty.

That's because seawater has salt dissolved in it.

Many types of rocks contain tiny **particles**, or pieces, of salt.

When waves crash against rocky cliffs, rock and salt get washed into the sea.

Rock and salt particles

Rivers make rocks break up and crumble.

Then rivers flow into the ocean, carrying tiny particles of rock and salt.

Ocean

River

If you look at salty seawater you can't see the salt. That's because it's dissolved.

Can we turn the salt back into solid particles?

You will need:
- A small jug
- Measuring spoons
- Water
- Salt
- A spoon
- A dark-coloured shallow dish
- A notebook and pencil

1) In the jug mix together 7 tablespoons of water and 2 teaspoons of salt.

2) Keep stirring with a spoon until the salt dissolves and the water is see-through again.

3) Pour the salty water into the shallow dish. Place the dish in a warm, sunny place, such as on a windowsill.

4) Keep checking the dish over the next few days.

Write down your results in your notebook. Can you include these science words in your results?

solid   **particles**

evaporate

liquid

What happens to the water?
What happens to the salt?

To find answers and more information, turn to page 29.

Water in Action

Every day, all around our Earth, water is changing from a liquid, to a gas or a solid, and back again. This is called the water cycle.

Let's see the water cycle in action!

The Sun shines on a puddle. The water evaporates and becomes water vapour.

The vapour floats up into the sky, where it is cold.

The vapour cools and condenses. It changes back into tiny drops of liquid water.

Many water drops join up and make a cloud. The drops stick together and form raindrops.

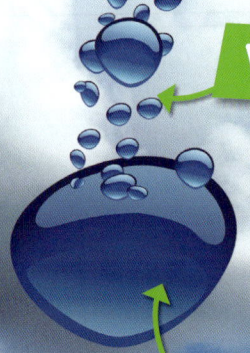

Water drops

Raindrop

If the air around a cloud is very cold, the water droplets freeze and become snowflakes.

The raindrops in a cloud grow bigger and heavier. Soon they are too heavy to stay up in the sky. Then they fall back to Earth — as rain!

Solid ice

Rain that falls on mountains may freeze and turn to solid ice. In summer, the ice might melt and flow down the mountain as a stream.

Rain lands on the ground. It might also land in an ocean, a lake, a river or a pond.

One day, that water may become vapour, and the cycle goes on.

Let's Talk Water

Did you enjoy being a water scientist? Let's check out some answers and discover some more cool things about water.

Page 5:
Where is all the water in your body hiding?
Your body is made of tiny parts called cells. Every cell in your body contains some water. Your blood is made of different types of cells and lots of water. Your urine (pee), saliva (spit), mucus (snot), sweat and tears all contain water.

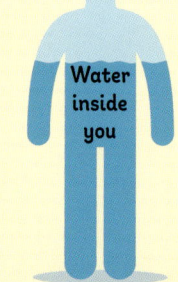
Water inside you

Page 7:
Here are some words that describe water:

wet	freezing
see-through	cold
sparkling	cool
fresh	warm
salty	hot
soggy	boiling
dripping	bubbly
wavy	frothy
runny	foamy

Page 9:
The colours in the hot water probably moved around and mixed together more than the colours in the cold water. Why?

When water is heated, the molecules move around faster than the molecules in cold water. The fast-moving, hot water molecules bumped the food colouring molecules many times. This made the coloured molecules mix and move.

Hot water

The water you poured was flowing. Being able to flow is a property of liquids.

Water does not have a shape. It takes the shape of the container it is in.

Water is a thin liquid. So is tea because it's made of water.
Water is thinner than ketchup, honey and shampoo.

Page 13:
The surface tension of the water made the pepper settle on top.

When your finger touched the pepper-covered water, did the pepper whiz to the edges of the water? The soap molecules on your finger broke the surface tension. The water molecules close to your finger were instantly pulled away by molecules that were not so close, and the grains of pepper went, too.

Page 15:
As you poured the water, did it flow down the string into the glass? Why did it do this?

The water from the small jug was sticking to itself and to the water on the wet string because of cohesion. It was also sticking to the string because of adhesion.

Page 17:
The warmth around the jars made the water evaporate. The water in the uncovered jar turned to water vapour. The vapour floated from the jar, so the level of the liquid water went down.

The water in the covered jar also evaporated. But once there was no sunshine, the jars cooled down. Then the vapour trapped in the covered jar turned back into liquid water (see pages 18–19), so the water level didn't change.

Page 19:
The glass containing the ice cubes has water on the outside. The ice cubes made the glass cold. Water vapour in the air touched the cold glass, cooled down and condensed back into liquid water. The same thing happens on a cold can.

The water on a cold glass or a cold can is known as condensation.

Page 21:
Did the ice cubes sprinkled with salt melt first? That's because salt is able to melt ice. It also makes it harder for water to freeze. Did your predictions match your results?

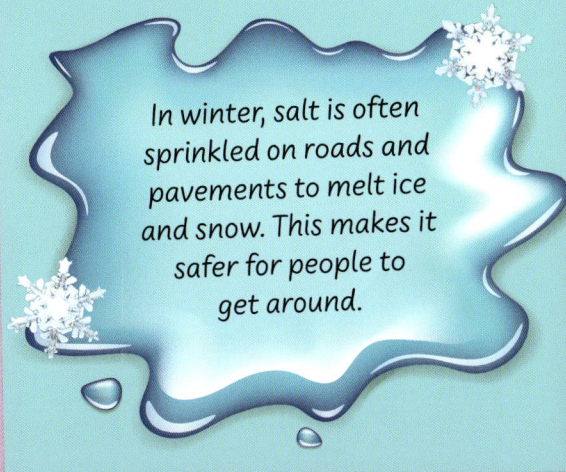

In winter, salt is often sprinkled on roads and pavements to melt ice and snow. This makes it safer for people to get around.

Page 25:
The sunshine warmed the liquid water in the dish and made it evaporate. After a few days, all that was left were the solid salt particles.

The dish's dark colour soaked up the Sun's heat and helped warm the water.

Glossary

adhesion
The property of water that allows it to stick to objects and surfaces, such as glass.

cohesion
The property of water that allows water molecules to stick to each other.

condense
To turn from a gas into a liquid. When water vapour cools down, it changes into liquid water.

dissolve
To become part of a liquid.

evaporate
To turn from a liquid to a gas. For example, when sunshine warms a puddle, the liquid water evaporates and changes into water vapour gas.

gas
A state of matter. Gases float in air and are neither solids nor liquids. Oxygen is an invisible gas in the air that people and animals need to breathe.

liquid
A state of matter. Liquids, such as water, flow and do not have a shape.

matter
All the real stuff around us such as water, clouds, rocks, books, animals and our bodies.

molecule
A group of atoms that are joined together. When two hydrogen atoms and one oxygen atom join, they make a water molecule.

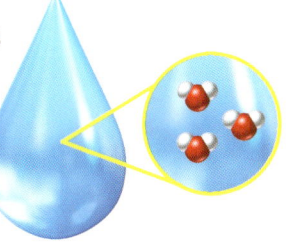

particles
Tiny parts of something, such as rock. Sand is made of rock particles.

property
A quality that helps describe what an object or substance is like. For example, a property of water is that it can be poured.

solid
An object with a definite size and shape. When water turns to ice, it becomes a solid.

surface tension
The sticking, or pulling together, of the outside layer of a liquid, such as water.

water vapour
The gas state of water. Water vapour is in the air all around us, but we can't see it because it's invisible.

Index

A
adhesion 14–15, 29

C
cohesion 10–11, 12, 14, 29
condensation 18–19, 26, 29

D
dissolving 22–23, 24–25

E
evaporation 16–17, 18, 25, 26, 29

F
fresh water 4–5, 26–27

G
gases 6, 16–17, 18–19, 20, 26–27, 29

I
ice 20–21, 27, 29

L
liquids 6–7, 16, 18–19, 20–21, 22, 25, 26–27, 28–29

M
matter 6, 20
molecules 8–9, 10, 12, 14, 28

S
seawater 4, 24–25, 27
solids 6, 20–21, 24–25, 26–27
surface tension 12–13, 28

W
water cycle 26–27
water vapour 16–17, 18–19, 20, 26–27, 29